PERENNIAL MATH

CONTEST PROBLEMS

2017-2018

Dr. Sylvia Dean was a gifted specialist and administrator for 26 years in Huntsville, Alabama. She created Perennial Math in 2011 upon completion of her doctoral degree. She has hosted many robotics, technology, and mathematics competitions throughout the years and has tirelessly worked on providing educational opportunities to students around the world. It has always been her ambition to leave students with the feeling that "math is cool."

These problems are challenging and students should know up front that scoring a 3 out of 6 on any of these tests is extraordinary and above average. Great problem-solvers learn persistence and appreciation for challenge. Good luck as you encourage your students to take the quest to become great mathletes.

Advanced
level

Grades 7-8

Name________________________________

PERENNIAL MATH

January 2017 Advanced Level

TEST #1

1) What is the least two-digit number that is twice the sum of its digits?

2) A recent favorite color poll revealed the following top 5 colors: Red (26%); Blue (12.9%); Green (4.8%); Orange (4.3%); and Purple (4%). If 10,000 people were interviewed, how many more people liked blue than green and orange?

3) The average of all 7 players on the basketball team was 13 points at the end of the game. However, the scoring average of everyone except for the point guard was 12. How many points did the point guard score?

4) In this figure all of the long sides are the same length and each one is twice as long as each of the shorter sides. The area of this figure is 200 cm^2. What is the perimeter of the figure?

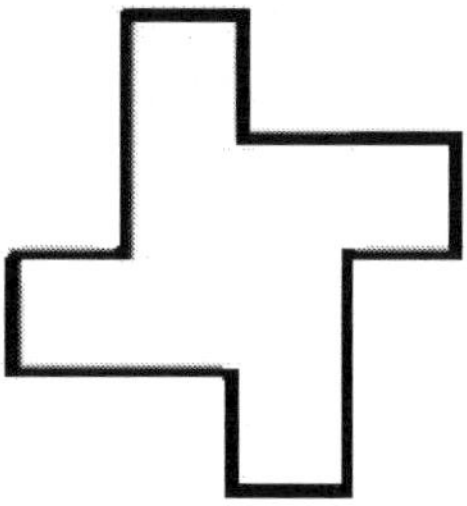

5) The letters in PERENNIAL and MATH are cycled as shown and placed in a numbered list. The next correct spelling of both words simultaneously will appear in what row number?

1.	PERENNIAL	MATH
2.	ERENNIALP	ATHM
3.	RENNIALPE	THMA

6) Morris has two containers. The first contains 31 percent alcohol and the second contains 25 percent alcohol. He receive an order for 18 liters of 29 percent alcohol. If he mixes alcohol from the two containers to fill this order, how many liters will come from the second container?

PERENNIAL MATH

TEST #1

January 2017 Advanced Level Answers

1) What is the least two-digit number that is twice the sum of its digits?	18
2) A recent favorite color poll revealed the following top 5 colors: Red (26%); Blue (12.9%); Green (4.8%); Orange (4.3%); and Purple (4%). If 10,000 people were interviewed, how many more people liked blue than green and orange?	380
3) The average of all 7 players on the basketball team was 13 points at the end of the game. However, the scoring average of everyone except for the point guard was 12. How many points did the point guard score?	19
4) In this figure all of the long sides are the same length and each one is twice as long as each of the shorter sides. The area of this figure is 200 cm^2. What is the perimeter of the figure? 	80cm
5) The letters in PERENNIAL and MATH are cycled as shown and placed in a numbered list. The next correct spelling of both words simultaneously will appear in what row number? 1. PERENNIAL MATH 2. ERENNIALP ATHM 3. RENNIALPE THMA	37
6) Morris has two containers. The first contains 31 percent alcohol and the second contains 25 percent alcohol. He receive an order for 18 liters of 29 percent alcohol. If he mixes alcohol from the two containers to fill this order, how many liters will come from the second container?	6

PERENNIAL MATH

TEST # 1

January 2017 Advanced Level Solutions

1) To find the least two-digit number that is twice the sum of its digits make a table.

tens digit	ones digit	sum	2x sum
1	1	2	4
1	2	3	6
1	3	4	8
1	4	5	10
1	5	6	12
1	6	7	14
1	7	8	16
1	8	9	18

18

2) A recent favorite color poll revealed the following top 5 colors: Red (26%); Blue (12.9%); Green (4.8%); Orange (4.3%); and Purple (4%). To solve for how many more people liked blue than green and orange of the 10,000 interviewed, solve for the number of people who liked blue, green, and orange then subtract the amounts. 12.9% of 10,000 = 1290. 4.8% of 10,000 = 480. 4.3% of 10,000=430. 1290 – (480 + 430) = 1290 – 910 = 380 people. Or 12.9% - (4.8% + 4.3%) = 3.8%. 3.8% of 10,000 = 380.

380

3) The average of all 7 players on the basketball team was 13 points at the end of the game. However, the scoring average of everyone except for the point guard was 12. To find out How many points the point guard scored, first find the total number of points scored. 7 x 13 = 91 points. Everyone except the point guard averaged 12 so their total points was 6 x 12 = 72 points. Subtract to see how many points the point guard scored. 91 – 72 = 19 points.

$$13 = \frac{x}{7} \qquad 12 = \frac{x}{6}$$

$$91 = x \qquad 72 = x \qquad 91 - 72 = 19$$

19

PERENNIAL MATH

TEST #1

January 2017 Advanced Level Solutions

4) The area of this figure is 200cm^2. To solve for the perimeter solve for the length of each side. You can divide the shape into 4 equal rectangles so 200cm^2/ 4 = 50cm^2. Since all of the long sides are the same length and each one is twice as long as each of the shorter sides, find the factors of 50cm^2 that are double each other: 5cm and 10cm. So now add up the sides of the figure. The perimeter of the figure is 80 cm.

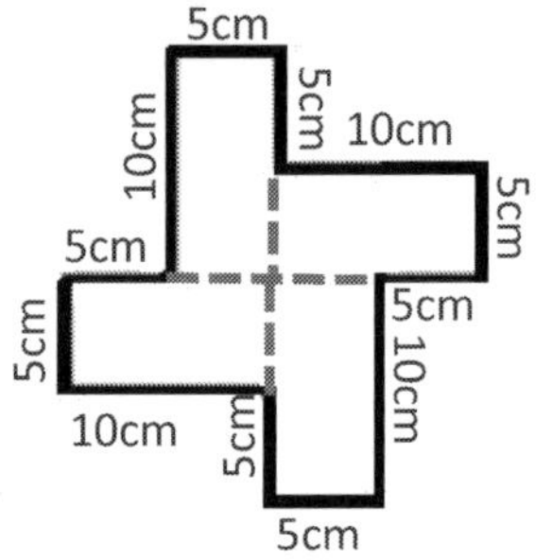

80cm

5) The word PERENNIAL appears every ninth row beginning with 1: 1, 10, 19, 28
The word MATH every fourth row beginning with 1: 1, 5, 9, 13…
If we use the least common denominator of 4 and 9, then the words PERENNIAL and MATH appear at the same time every 36 rows. Since they appear in row 1, then the next row they will appear simultaneously is row 37.

37

6) Let *x* stand for the number of liters of 31% solution, and let *y* stand for the number of liters of 25% solution.

	Amount of container (liters)	% gas	Amount of gas (liters)
Container X	x	0.31	.31x
Container Y	18-x	0.25	.25(18-x)
Together	18	0.29	.29(18)

0.31x + 0.25(18 – x) = 0.29(18)
0.31x + 4.5 - 0.25x = 5.22
0.06x = 0.72
x = 12
Now solve for the amount in Container Y. 18 – 12 = 6 liters.

6

Name______________________________

PERENNIAL MATH

February 2017 Advanced Level

TEST #2

1) How many two-digit numbers between 1-100 have the units digit greater than the tens digit?	
2) 5/6 of a number is 2500. What is 2/3 of that number?	
3) Suppose 90 days ago was Saturday. What day of the week will it be 90 days from today?	
4) ABCD is a square with area 144m². E, F, G, and H are points on sides AB, BC, CD, and DA, respectively, such that AH = BE = CF = DG = 2m. What is the area of the shaded region? 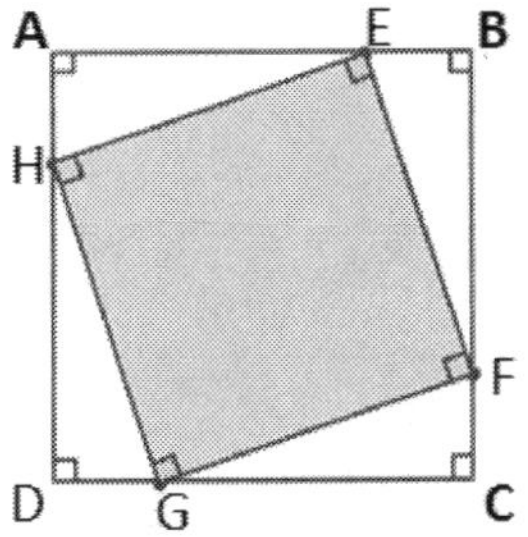	
5) 9 green balls weigh as much as 4 black balls. 8 black balls weigh as much as 15 orange balls. 10 orange balls weigh as much as 27 red balls. How many red balls weigh the same as 4 green balls?	
6) A basketball team has won seven of its first fifteen games. How many wins in their next ten games will raise their winning percentage for the season to 60%?	

PERENNIAL MATH

TEST # 2

February 2017 Advanced Level Answers

1) How many two-digit numbers between 1-100 have the units digit greater than the tens digit?	36
2) 5/6 of a number is 2500. What is 2/3 of that number?	2000
3) Suppose 90 days ago was Saturday. What day of the week will it be 90 days from today?	Thursday
4) ABCD is a square with area 144m². E, F, G, and H are points on sides AB, BC, CD, and DA, respectively, such that AH = BE = CF = DG = 2m. What is the area of the shaded region? 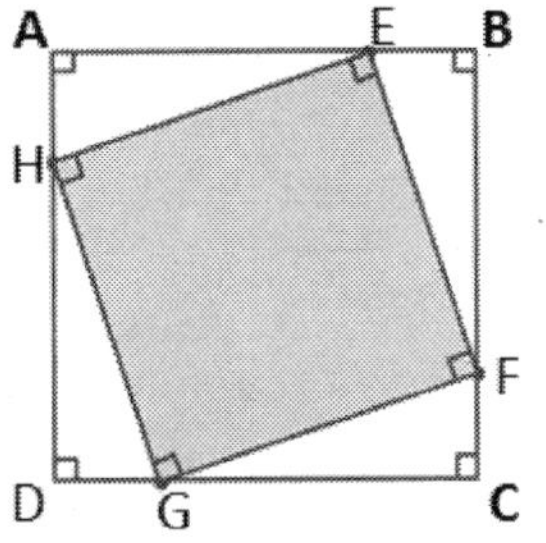	104
5) 9 green balls weigh as much as 4 black balls. 8 black balls weigh as much as 15 orange balls. 10 orange balls weigh as much as 27 red balls. How many red balls weigh the same as 4 green balls?	9
6) A basketball team has won seven of its first fifteen games. How many wins in their next ten games will raise their winning percentage for the season to 60%?	8

PERENNIAL MATH TEST #2

February 2017 Advanced Level Solutions

1) Make a list to find out how many two-digit numbers between 1-100 have the units digit greater than the tens digit.

Tens	Units								How many Numbers	
1	2	3	4	5	6	7	8	9	8	
2	3	4	5	6	7	8	9		7	
3	4	5	6	7	8	9			6	
4	5	6	7	8	9				5	
5	6	7	8	9					4	
6	7	8	9						3	
7	8	9							2	
8	9								1	
									36	Total

36

2) 5/6 of a number is 2500. To find 2/3 of that number use algebra. First solve for the original number n and then find 2/3n.

5/6n = 2500

n = 2500 * 6/5

n = 15,000/5

n = 3000

Now find 2/3 of 3000. The answer is 2000.

2000

3) Suppose 90 days ago was Saturday. What day of the week will it be 90 days from today? 90 days before today is equal to 90 ÷ 7 = 12r6. So 90 after a Saturday is Friday- today is Friday. Now 90 days after today (Friday) is Thursday.

Thursday

PERENNIAL MATH

TEST # 2

February 2017 Advanced Level Solutions

4) The area of square ABCD is 144m^2. Therefore the side lengths are 12m since 12 x 12 = 144 and the area = length x width. Each of the smaller parts of the sides of square ABCD are 2m so the longer parts of the sides are 12 - 2 = 10m.

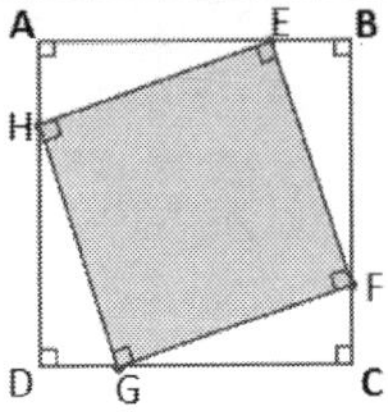

In right ΔAEH, AH= 2 and AE = 8. We can use one side as the base and the other as the height in the calculation of the area of the triangle since they are perpendicular to each other.

Therefore the area of ΔAEH = $\frac{AE\ x\ AH}{2} = \frac{10\ x\ 2}{2} = 10m^2$

Since each of the triangles has the same base length and height, their areas are equal and the total area of the four triangles is 4 x 10 m^2 = 40 m^2.

The area of the shaded region is a square EFGH and can be determined by subtracting the area of the four triangles from the area of square ABCD.

Therefore the area of the shaded region square EFGH = 144 - 40 = 104m^2.

104

5) 9 green balls weigh as much as 4 black balls. 8 black balls weigh as much as 15 orange balls. 10 orange balls weigh as much as 27 red balls. To solve for how many red balls weigh the same as 4 green balls make a conversion chart. Since 4 black balls are equal to 9 green balls, then 8 black balls are equal to 18 green balls. Therefore, 18 green balls weigh the same as 15 orange balls which can be reduced to 6 green balls weighing the same as 5 orange balls. Seeing that 10 orange balls weigh as much as 27 red balls, then 12 green balls weigh the same as 27 red balls. The ratio of green balls to red is 12:27 which can be reduced to 4:9 Therefore, 9 red balls are equal to 4 green balls.

Green	Black	Orange	Red
9	4		
18	8	15	
12		10	27
4			?

9

6) A basketball team has won seven of its first fifteen games. How many wins in their next ten games will raise their winning percentage for the season to 60%? First solve for the number of games they must win for the whole season to have 60%. 15 games plus 10 more games is a total of 25 games. 60% of 25 would be 15 games. Since the team has won only seven games, they must win 8 of the remaining 10 games to have 60% winning percentage.

8

Name________________________________

PERENNIAL MATH

TEST #3

March 2017 Advanced Level

1) $x = 7 \cdot 10^5 + 4 \cdot 10^3 + 3 \cdot 10^2 + 6 \cdot 10^0$. What is x?	
2) A toy maker built some three-wheeled toys and some four-wheeled toys. When he finished he saw that there were three more three-wheeled toys than four-wheeled toys and that he had used 72 wheels. How many three-wheeled toys did he build?	
3) Ellis works at a the tennis store. Her manager asked her to make a display using 5 different tennis dresses from 8 available styles. How many ways can Ellis choose to display the dresses?	
4) Given that the sides of the square are equal to 16 cm, what is the area of the shaded region? *(Assume π = 3.14)* 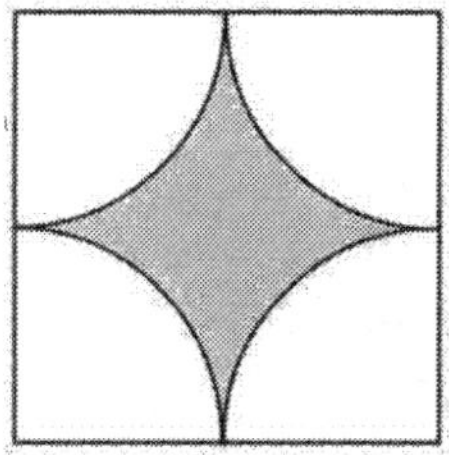	
5) On a scale, 3 apples and 1 banana balances 10 plums. One apple and 6 plums balance 1 banana. How many plums will balance one banana?	
6) The sequence 1, 2, 3, 4, 5, 7, 8… consists of numbers that are perfect squares, prime, and perfect cubes. What will be the 50th term in this sequence?	

PERENNIAL MATH

TEST #3

March 2017 Advanced Level Answers

1) $x = 7 \cdot 10^5 + 4 \cdot 10^3 + 3 \cdot 10^2 + 6 \cdot 10^0$. What is x?	704,306
2) A toy maker built some three-wheeled toys and some four-wheeled toys. When he finished he saw that there were three more three-wheeled toys than four-wheeled toys and that he had used 72 wheels. How many three-wheeled toys did he build?	12
3) Ellis works at a the tennis store. Her manager asked her to make a display using 5 different tennis dresses from 8 available styles. How many ways can Ellis choose to display the dresses?	6,720
4) Given that the sides of the square are equal to 16 cm, what is the area of the shaded region? *(Assume π = 3.14)* 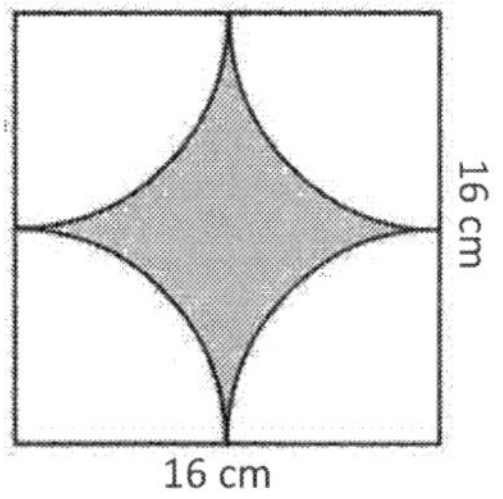	55.04
5) On a scale, 3 apples and 1 banana balances 10 plums. One apple and 6 plums balance 1 banana. How many plums will balance one banana?	7
6) The sequence 1, 2, 3, 4, 5, 7, 8... consists of numbers that are perfect squares, prime, and perfect cubes. What will be the 50th term in this sequence?	149

PERENNIAL MATH

TEST # 3

March 2017 Advanced Level Solutions

1) $x = 7 \cdot 10^5 + 4 \cdot 10^3 + 3 \cdot 10^2 + 6 \cdot 10^0$

$x = 7 \cdot 100000 + 4 \cdot 1000 + 3 \cdot 100 + 6 \cdot 1$

$x = 700000 + 4000 + 300 + 6$

$x = 704{,}306$

704,306

2) A toy maker built some three-wheeled toys and some four-wheeled toys. When he finished he saw that there were three more three-wheeled toys than four-wheeled toys and that he had used 72 wheels. To find out how many three-wheeled toys he built make a table. There were 12 three-wheeled toys.

3 wheeled toys	4	5	6	7	8	9	10	11	12
4 wheeled toys	1	2	3	4	5	6	7	8	9
3 wheeled toy wheels	12	15	18	21	24	27	30	33	36
4 wheeled toy wheels	4	8	12	16	20	24	28	32	36
Total wheels	16	23	30	37	44	51	58	65	72

12

3) Ellis works at a the tennis store. Her manager asked her to make a display using 5 different tennis dresses from 8 available styles. How many ways can Ellis display the dresses (order is not important)?

A permutation is an arrangement of all or part of a set of objects, *with* regard to the order of the arrangement. This means that XYZ is considered a different permutation than ZYX. The number of permutations of *n* objects taken *r* at a time is denoted by ${}_nP_r$.

Thus she has ${}_8P_5 = 8! / (8 - 5)!$ or $8! / 3!$

$= 8 \times 7 \times 6 \times 5 \times 4$

$= 6{,}720$

6,720

PERENNIAL MATH

TEST # 3

March 2017 Advanced Level Solutions

4) To find the area of the shaded region, first find the area of the entire square and then subtract the area of the circle (white regions). Since the sides are 16 cm, the area of the square is 16 cm x 16 cm = 256 cm^2. The midpoint of the side of the square is equal to a radius of the circle that can be formed by the four white regions. That would make the area of the circle (4 white regions) be $8cm^2 * \pi = 200.96cm^2$. Now subtract $256cm^2 - 200.96cm^2 = 55.04\ cm^2$.

55.04

5) On a scale, 3 apples and 1 banana balances 10 plums. One apple and 6 plums balance 1 banana. How many plums will balance one banana?

3A + B = 10P

A + 6P = B (substitute this equation in the previous equation for the banana)

3A + (A + 6P) = 10P

4A + 6P = 10P (subtract 6P from both sides)

4A = 4P

Therefore 1 A weighs the same as 1 P. That would mean that 7 plums would balance 1 banana.

7

6) The sequence 1, 2, 3, 4, 5, 7, 8… consists of numbers that are perfect squares, prime, and perfect cubes. Make a list to find the 50th term in this sequence. You could break it down and make lists of primes, squares and cubes so that you don't omit one.

Squares: 1, 4, 9, 16, 25, 36, 49, 64, 81, 100, 121, 144

Primes: 2, 3, 5, 7, 11, 13, 17, 19, 23, 29, 31, 37, 41, 43, 47, 53, 59, 61, 67, 71 73, 79, 83, 89, 97, 101, 103, 107, 109, 113, 127, 131, 137, 139, 149

Cubes: 1, 8, 27, 64, 125, 216

1, 2, 3, 4, 5, 7, 8, 9, 11, 13,
16, 17, 19, 23, 25, 27, 29, 31, 36, 37,
41, 43, 47, 49, 53, 59, 61, 64, 67, 71,
73, 79, 81, 83, 89, 97, 100, 101, 103, 107,
109, 113, 121, 125, 127, 131, 137,139, 144, 149

149

Name________________________________

PERENNIAL MATH

TEST #4

April 2017 Advanced Level

1) If 4 and x have the same mean as 3, 6, and 21, what is the value of x?

2) A tractor's fuel tank has 20 gallons of a 80% fuel mixture (80% gasoline and 20% oil). The manual recommends a mixture of gasoline and oil where there is 60% gasoline. How many gallons of a 50% fuel mixture must be added to make the recommended mixture of 60% gasoline?

3) What is the units digit of the integer 3^{2014}?

4) The sum of the symbols of each row the square are given. What is the value of 1 pentagon?

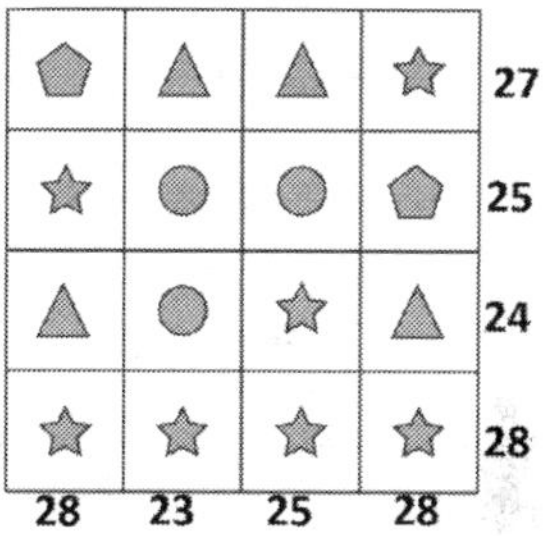

5) The sum of the lengths of three sides of a rectangle is 45 cm. Each side is a whole number of cm in length. The length is 12 cm more than the width. Find the area of the rectangle in square cm.

6) What is the greatest prime factor of $9^{18} - 3^{32}$?

PERENNIAL MATH TEST #4

April 2017 Advanced Level Answers

1) If 4 and x have the same mean as 3, 6, and 21, what is the value of x?	16
2) A tractor's fuel tank has 20 gallons of a 80% fuel mixture (80% gasoline and 20% oil). The manual recommends a mixture of gasoline and oil where there is 60% gasoline. How many gallons of a 50% fuel mixture must be added to make the recommended mixture of 60% gasoline?	40
3) What is the units digit of the integer 3^{2014}?	9
4) The sum of the symbols of each row the square are given. What is the value of 1 pentagon? 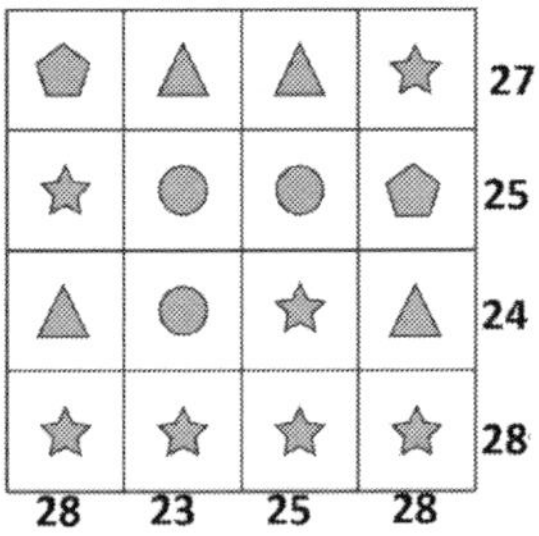	8
5) The sum of the lengths of three sides of a rectangle is 45 cm. Each side is a whole number of cm in length. The length is 12 cm more than the width. Find the area of the rectangle in square cm.	253 or 133
6) What is the greatest prime factor of $9^{18} - 3^{32}$?	5

PERENNIAL MATH

TEST # 4

April 2017 Advanced Level Solutions

1) If 4 and x have the same mean as 3, 6, and 21, what is the value of x? Find the mean of 3, 6, and 21. 3 + 6 + 21 = 30. 30/3 = 10. (4 + x) must have a mean of 2. So their sum must be 20. x = 16.

16

2) Make a table with three columns: Total Gallons of Mixture, % Gas Mixture and Total Amount of Gas in Mixture. Fill the table out with the knowns and solve for the unknown (x), the total amount of the 50% mixture to added.

	Total Gallons	% Gas Mixture	Total Amout of Gas in Mixture
Starting Mixture	20	80%	16
Mixture to Add	x	50%	0.5x
Final Mixture	20 + x	60%	

Now find the Total Amount of Gas in the Final Mixture (shaded cell above) in terms of "x" . This can be done two ways: Total Gallons of Mixture times the % Gas Mixture (row) and also by adding the two Total Amount of Gas in Mixture amounts (column). If you set those two equations equal to each other we can now solve for our unknown "x", the total amount of the 50% mixture to add.

$0.6\,(20 + x) = 16 + 0.5x$

$12 + 0.6x = 16 + 0.5x$

$0.6x = 4 + 0.5x$

$0.1x = 4$

$x = 40$ gallons

40

3) What is the units digit of the integer 3^{2014}?

A unit digit pattern will emerge if you do the multiplication:

3^1 = <u>3</u>

3 x 3 = <u>9</u>

9 x 3 = 2<u>7</u>

27 x 3 = 8<u>1</u>

81 x 3 = 24<u>3</u>

So the pattern is 9, 7, 1, 3 and the pattern repeats after every 4 numbers. Divide 2014 by 4 and the answer is 503 with remainder of 2. So the answer would end with 9 in the units digit.

9

PERENNIAL MATH

TEST #4

April 2017 Advanced Level Solutions

4) The value of a star is 28/4 = 7. Using the 3rd column, you can deduce that a Circle + Triangle = 11 because we can subtract the value of 2 Stars from 25. Now solve for the value of a Circle in column 2 because we know that the value of a Circle and Triangle is 11 and the value of a Star is 7, so the remaining circle is worth 23 – (11 + 7). The Circle is worth 5. Now we can solve for the value of the Pentagon by plugging in the values of the Circles and Star in the second row. 25 – (7 + 5 + 5) = 8. The Pentagon is worth 8.

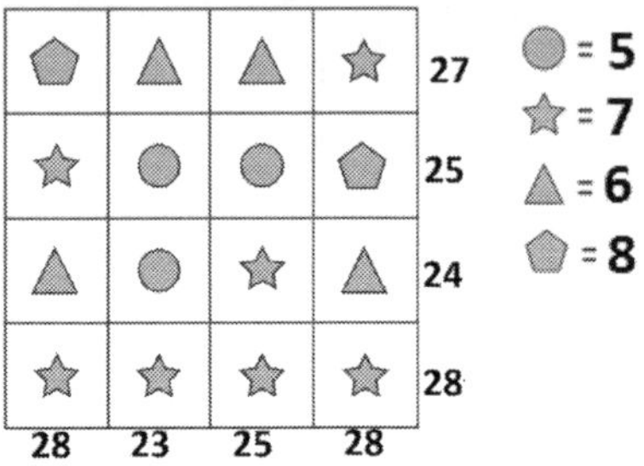

8

5) Make a table of the possible lengths and widths of the triangle with the length being 12 more than the width. Then make a column that adds 2 widths + 1 length. Two widths of 11 and a length of 23 will yield 45. Since the length and width have been found you can multiply the width by the length to find the area of the rectangle. The area of the rectangle is (11 x 23) or 253 cm^2. An additional solution is 2 lengths + 1 width = 45. 2(19) + 7 = 45. 19 x 7 = 133 cm^2.

Width (W)	Length (L) W + 12	2W + L
1	13	15
2	14	18
3	15	21
4	16	24
5	17	27
6	18	30
7	19	33
8	20	36
9	21	39
10	22	42
11	23	45

253 or 133

6) $9^{18} - 3^{32} = 3^{36} - 3^{32} = 3^{32} \times (3^4 - 1) = 3^{32} \times 80 = 3^{32} \times 2^4 \times 5$

$4^{18} - 2^{32} = 2^{36} - 2^{32} = 2^{32} \times (2^4 - 1) = 2^{32} \times 15 = 2^{32} \times 3 \times 5$

5

Name________________________________

PERENNIAL MATH

TEST #1

November 2017 Advanced Level

1) Find the sum of the exponents in the prime factorization of 80.	
2) Two thirds of Brandon's coins are quarters; the rest are dimes. The coins are worth a total of $12. How many quarters does Brandon have?	
3) One day, William goes to the mall and spends 1/2 his money on a pair of jeans. He spends 1/3 of what he has left on a CD and then 1/4 of the remainder of that on lunch. If he returns home with $12.00, how much did he start with?	
4) If you start at Point X and can only move in the gray striped pathways without crossing or touching a corner of your path again, how many different ways can you go from Point X to Point Y? 	
5) If simple interest on a credit card bill for $6000 for 3.5 years is $945, find the simple interest on a $5000 for 8 months at the same rate.	
6) The sum of the first 25 multiples of 5 is: 5 + 10 + 15 + ... + 125. The sum of the first 25 multiples of 6 is: 6 + 12 + 18 +... + 150. What is the difference of the two sums?	

PERENNIAL MATH

TEST # 1

November 2017 Advanced Level Answers

1) Find the sum of the exponents in the prime factorization of 80.	5
2) Two thirds of Brandon's coins are quarters; the rest are dimes. The coins are worth a total of $12. How many quarters does Brandon have?	40
3) One day, William goes to the mall and spends 1/2 his money on a pair of jeans. He spends 1/3 of what he has left on a CD and then 1/4 of the remainder of that on lunch. If he returns home with $12.00, how much did he start with?	48
4) If you start at Point X and can only move in the gray striped pathways without crossing or touching a corner of your path again, how many different ways can you go from Point X to Point Y? 	16
5) If simple interest on a credit card bill for $6000 for 3.5 years is $945, find the simple interest on a $5000 for 8 months at the same rate.	150
6) The sum of the first 25 multiples of 5 is: 5 + 10 + 15 + … + 125. The sum of the first 25 multiples of 6 is: 6 + 12 + 18 +… + 150. What is the difference of the two sums?	325

PERENNIAL MATH

TEST # 1

November 2017 Advanced Level Solutions

1) Prime Factors of 80 = 2, 2, 2, 2, 5 which is the same as $2^4 \times 5$. The exponents are 4 and 1. So 4 + 1 = 5

5

2) Two thirds of Brandon's coins are quarters; the rest are dimes and are worth a total of $12. To find how many quarters Brandon has make a chart. You will discover that If he has 10 quarters the total sum is $3. If you were to multiply that by 4 times the amount, then the total would be $12. So he has 10 x 4 = 40 quarters and 5 x 4 = 20 dimes.

Quarters	Dimes	Quarters amount	Dimes amount	Sum
2	1	0.5	0.1	0.6
4	2	1	0.2	1.2
6	3	1.5	0.3	1.8
8	4	2	0.4	2.4
10	5	2.5	0.5	3

40

3) Draw a picture. William spends 1/2 his money on a pair of jeans so mark out 1/2. He spends 1/3 of what he has left on a CD so mark out 1/3 of the remaining half. Finally he spent 1/4 of the remainder on lunch so mark out 1/4 of the remaining money. He returns home with $12.00, so there are 3 parts left and each part is worth $4. Now you can solve for what the half of the money is. It is worth $24. So he started with $48. You can work this out forwards to check. $48/2 = $24. Secondly he spends 1/3 which is $8 and is left with $16. He finally spends 1/4 or $4 of this and is left with $12.

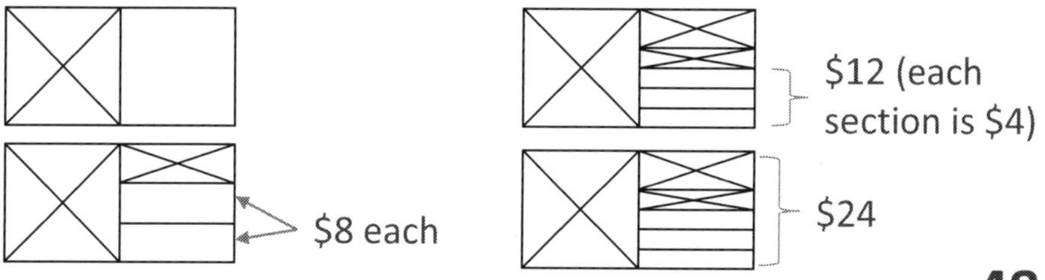

48

PERENNIAL MATH

TEST # 1

November 2017 Advanced Level Solutions

4)

16

5) If simple interest on a credit card bill for $6000 for 3.5 years is $945, find the simple interest on a $5000 bill for 8 months at the same rate. Figure the rate of the 1st investment.

Interest = Principle ▪ rate ▪ time

945 = 6000 ▪ r ▪ 3.5

r = 945/[6000 ▪ 3.5]

r = 0.045 or 4.5% per year

Figure the interest on the 2nd investment using the rate that was discovered in the first scenario:

Interest = Principle ▪ rate ▪ time

I = 5000 ▪ 0.045 ▪ (2/3)

I = $150

150

6) The sum of the first 25 multiples of 5 is: 5 + 10 + 15 + ... + 125.
The sum of the first 25 multiples of 6 is: 6 + 12 + 18 +... + 150. If you were to line up the multiples and do subtraction, you will notice that there would be 1 + 2 + 3 + 4 ... + 25. So now you can use smart addition to do the addition and 1 + 2 + 3...+ 25. If you were to group the sums in pairs such as 1 + 24, 2 + 23, 3 + 22, ... and then also add the final number 25 (see below), you would have 12 pairs that add up to 25 along with the extra 25. 13 x 25 = 325. Therefore, the difference of the two sums of the first 25 multiples of 5 and 6 is 325.

Sum of Multiples of 6= 6 + 12 + 18 + 24 + 30 + 36...
Sum of Multiples of 5= <u>5 + 10 + 15 + 20 + 25 + 30...</u>
Difference 1 + 2 + 3 + 4 + 5 + 6...+ 25 = 325

325

Name________________________________

PERENNIAL MATH

December 2017 Advanced Level

TEST #2

1) 789 x 2750 = 14465 x _____	
2) Marvin and Gary have 29 marbles together. Gary and Brian have 37 marbles together. Marvin and Brian have 32 marbles together. How many marbles does Gary have by himself?	
3) Trish is 1/6 the age of Carolyn this year. Next year she will be 1/5 the age of Carolyn. In how many years will she be 1/3 Carolyn's age?	
4) In this cryptarithm, the letters A, B, and C stand for different digits. What is the largest three-digit number that the letters CBA can represent? $$\begin{array}{r} AB \\ \times\ \ B \\ \hline CBA \end{array}$$	
5) Of 168 reviews of a Darby's restaurant, everyone liked something. 96 people liked the food, 64 people liked the service, and 56 people liked the atmosphere. 20 people liked the food and service, 17 people liked the food and atmosphere, and 16 people liked the service and atmosphere. Five people liked the food, service, and atmosphere. How many people liked the atmosphere, but neither the food or service?	
6) A statue is made up of only copper and gold. The ratio of the weights of copper to gold is 3 to 17. If the weight of the statue is 55 pounds, what is the value of the gold in the statue if gold is worth $1200 per ounce?	

PERENNIAL MATH

TEST #2

December 2017 Advanced Level Answers

1) 789 x 2750 = 14465 x _____	150
2) Marvin and Gary have 29 marbles together. Gary and Brian have 37 marbles together. Marvin and Brian have 32 marbles together. How many marbles does Gary have by himself?	17
3) Trish is 1/6 the age of Carolyn this year. Next year she will be 1/5 the age of Carolyn. In how many years will she be 1/3 Carolyn's age?	6
4) In this cryptarithm, the letters A, B, and C stand for different digits. What is the largest three-digit number that the letters CBA can represent? $$\begin{array}{r} AB \\ \times\ \ B \\ \hline CBA \end{array}$$	679
5) Of 168 reviews of a Darby's restaurant, everyone liked something. 96 people liked the food, 64 people liked the service, and 56 people liked the atmosphere. 20 people liked the food and service, 17 people liked the food and atmosphere, and 16 people liked the service and atmosphere. Five people liked the food, service, and atmosphere. How many people liked the atmosphere, but neither the food or service?	28
6) A statue is made up of only copper and gold. The ratio of the weights of copper to gold is 3 to 17. If the weight of the statue is 55 pounds, what is the value of the gold in the statue if gold is worth $1200 per ounce?	897,600

PERENNIAL MATH

TEST #2

December 2017 Advanced Level Solutions

1) 789 x 2750 = 14465 x _____

789 x 2750 = 2,169,750

2,169,750 ÷ 14,465 = 150

150

2) Use algebra.

$$\begin{array}{lr} M + G = 29 & (1) \\ G + B = 37 & (2) \\ + \underline{M + B = 32} & (3) \end{array}$$

2M + 2G + 2B = 98 divide both sides by 2 and M + G + B = 49.

Use M + G + B = 49 and subtract equation (3) from above to solve for G.

$$\begin{array}{l} M + G + B = 49 \\ - \underline{M + B = 32} \\ G = 17 \end{array}$$

So, Gary had 17 marbles.

17

3) Trish is 1/6 the age of Carolyn this year. Next year she will be 1/5 the age of Carolyn. To solve for how many years until she will be 1/3 Carolyn's age use algebra:

If Trish is x and Carolyn is y, then

y = 6x (this year)

y + 1 = 5(x+1) (next year)

So for example, for this year if Trish is 4, then Carolyn is 24

Now we ask when

24 + n = 3(4+n)

n = 6

So, in 6 years, Trish will be 10, and Carolyn will be 30

Or make a chart:

Trish	1	2	3	4
Carolyn	6	12	18	24
Trish + 1	2	3	4	5
Carolyn + 1	7	13	19	25

Trish	4	5	6	7	8	9	10
Carolyn	24	25	26	27	28	29	30

6

PERENNIAL MATH

TEST #2

December 2017 Advanced Level Solutions

4) For this problem we will focus on solving for A and B first. B cannot be a 2 because the product would only be a two-digit number. B also cannot 0, 1, 5, or 6 because they yield the same number (B). With trial and error you will discover that B can be 7 or 8. Then the products that are possible are 384, and 679. The largest possible product is 97 x 7 = 679.

$$\begin{array}{r} AB \\ \times\ \ B \\ \hline CBA \end{array}$$

679

5) Use a Venn diagram and label each part. All 7 Regions add up to 168. Food = I + II + III + V = 96; Atmosphere = II + IV + V + VII = 56; Service = III + V + VI + VII = 64. We need to solve for Region IV (people that liked the atmosphere, but neither the food or service). Label Region V with 5 to start with and use it to solve for the other Regions. Since 20 people liked food and service, then Region III = 20 - 5 or 15. Since 17 people liked food and atmosphere, then Region II = 17 – 5 or 12. Since 16 people liked service and atmosphere then Region VII = 16 – 5 or 11. Now to solve for how many people liked atmosphere, but neither food or service, we can subtract Regions II, V, and VII from (56). 56 - 12 - 5 - 11 = 28. Twenty-eight people liked atmosphere only. If you want to check your work you can solve for Regions I and VI and see if the up all the regions add up to 168.

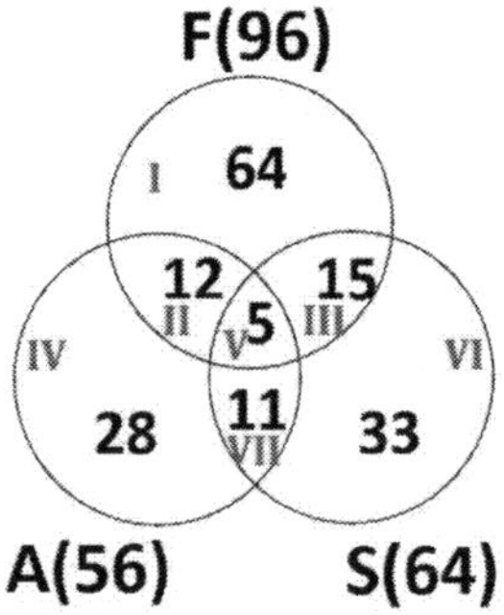

28

6) This is a mixture problem. Note that there are 20 parts to the statue: 3 + 17 Convert the statue's total weight from pounds to ounces: 55 pounds x 16 ounces per pound = 880 ounces. Next, calculate the how many ounces per part: 880 ounces ÷ 20 parts = 44 20-ounce parts to the statue. We know that there are 17 ounces of gold in each of the 44 parts so calculate the total ounces of gold in the statue: 44 x 17 ounces = 748 ounces of gold. Now determine the total value of the statue's gold: 748 ounces of gold x $1200 per ounce = $897,600.

897,600

Name________________________________

PERENNIAL MATH

TEST #3

January 2018 Advanced Level

1) 60 is 75% of what number?	
2) The ratio of blue bathing suits to red bathing suits in the store is 7:5. If there are 24 more blue bathing suits than red bathing suits, how many blue and red bathing suits are in the store?	
3) The number 28 can be expressed as the sum of two prime numbers in two different ways. What is the sum of the smallest two of those four primes?	
4) Find the height h of the trapezoid so that its area is equal to 60 cm^2. 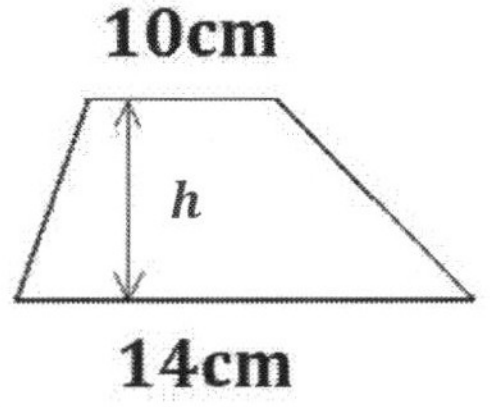	
5) Marla earned the following scores on seven math tests: 88, 95, 84, 100, 86, 99, and 94. If Marla's lowest grade is dropped, what will she have to score on the next test to have a 94 average?	
6) If a couple has 7 children, what is the probability that they will be all one gender (either all girls or all boys)? Express answer as a fraction.	

PERENNIAL MATH

TEST # 3

January 2018 Advanced Level Answers

1) 60 is 75% of what number?	80
2) The ratio of blue bathing suits to red bathing suits in the store is 7:5. If there are 24 more blue bathing suits than red bathing suits, how many blue and red bathing suits are in the store?	144
3) The number 28 can be expressed as the sum of two prime numbers in two different ways. What is the sum of the smallest two of those four primes?	16
4) Find the height h of the trapezoid so that its area is equal to 60 cm^2. 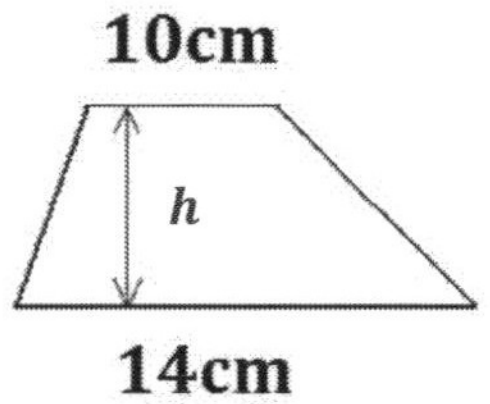	5
5) Marla earned the following scores on seven math tests: 88, 95, 84, 100, 86, 99, and 94. If Marla's lowest grade is dropped, what will she have to score on the next test to have a 94 average?	96
6) If a couple has 7 children, what is the probability that they will be all one gender (either all girls or all boys)? Express answer as a fraction.	1/64

PERENNIAL MATH

TEST #3

January 2018 Advanced Level Solutions

1) 60 is 75% of x translates to

60 = 0.75 * x or 60 = (3/4)x

Solving for x,

x = (4/3)*60

x = 80

80

2) The ratio of blue bathing suits to red bathing suits in the store is 7:5. If there are 24 more blue bathing suits than red bathing suits, how many blue and red bathing suits are in the store? Note that there are 7 + 5 = 12 total portions and we can see that there are two more portions of blue than red. Since we know that there are 24 more blue suits than red, we can solve that 24/2 = 12 and that each portion is 12 bathing suits. Since there are 12 portions of suits, 12 x 12 = 144 total red and blue bathing suits.

144

3) The number 28 can be expressed as the sum of two prime numbers in two different ways. To find the sum of the smallest two of the four primes, first find the two sums that are possible. Make a list of the primes and see which sums work: 2, 3, 5, 7, 11, 13, 17, 23. The primes that make 28 are 5 + 23 and 11 + 17. So the sum of those two smallest primes are 5 + 11 = 16.

16

PERENNIAL MATH

TEST #3

January 2018 Advanced Level Solutions

4) Use the following formula for an area of a regular polygon to find the height.

$$A = \frac{1}{2} \times (b_1 + b_2) \times h$$

$60cm^2 = ½ \times (10cm + 14cm) \times h$
$60cm^2 = ½ \times (24cm) \times h$
$60cm^2 = 12cm \times h$
$5cm = h$

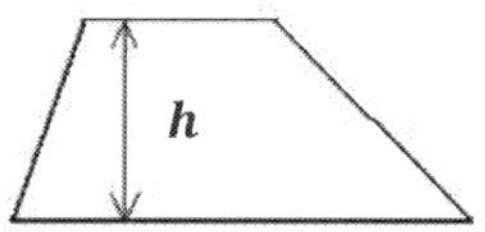

5

5) Marla earned the following scores on seven math tests: 88, 95, 84, 100, 86, 99, and 94. If Marla's lowest grade is dropped, she will have to score (94 x 7) = 658 total points to average 94. She currently has a total of (88 + 95 + 100 + 86 + 99 + 94) or 562 points when the 84 is dropped. So in order to find out what she will have to score on the next test to get a 94 average you will subtract her current points from the total points that are needed. 658 – 562 = 96. She will need to score a 96 on the next test.

96

6) If a couple has 7 children, what is the probability that they will be all one gender? After the first child is born, all children must match that gender. Each of the next six births there is a 1 in 2 probability that a matching gender child will be born. Six births: ½ x ½ x ½ x ½ x ½ x ½ = 1/64

1/64

Name________________________________

PERENNIAL MATH

TEST #4

February 2018 Advanced Level

1) What is the smallest positive integer with exactly 8 positive factors?	
2) Twice a number is 21 more than half the number. What is the number?	
3) A carpenter built some decorative tables. Some had three legs and some had 4 legs. When he finished he counted 49 tables and 169 legs. How many 4-legged tables did he build?	
4) If you stack balls in a pyramid with 1 on the first layer, 4 on the second layer, 9 on the third layer, and 16 on the fourth layer, how many total balls will you use to make a pyramid with 12 layers? 	
5) The school spirit bus traveled 150 miles to the coliseum at a rate of 50 mph because the radiator was leaking. The bus was repaired during the game and they returned home on the same road. If their average rate was 60 mph, how fast did they travel on the trip back home?	
6) Eric gave Craig and Trevor as many marbles as each had. Craig then gave Eric and Trevor as many marbles as each had. Then Trevor gave Eric and Craig as many marbles as each had and then they all had 16. How many marbles did Eric have to begin with?	

PERENNIAL MATH

TEST #4

February 2018 Advanced Level Answers

1) What is the smallest positive integer with exactly 8 positive factors?	24
2) Twice a number is 21 more than half the number. What is the number?	14
3) A carpenter built some decorative tables. Some had three legs and some had 4 legs. When he finished he counted 49 tables and 169 legs. How many 4-legged tables did he build?	22
4) If you stack balls in a pyramid with 1 on the first layer, 4 on the second layer, 9 on the third layer, and 16 on the fourth layer, how many total balls will you use to make a pyramid with 12 layers? 	650
5) The school spirit bus traveled 150 miles to the coliseum at a rate of 50 mph because the radiator was leaking. The bus was repaired during the game and they returned home on the same road. If their average rate was 60 mph, how fast did they travel on the trip back home?	75
6) Eric gave Craig and Trevor as many marbles as each had. Craig then gave Eric and Trevor as many marbles as each had. Then Trevor gave Eric and Craig as many marbles as each had and then they all had 16. How many marbles did Eric have to begin with?	26

PERENNIAL MATH

TEST # 4

February 2018 Advanced Level Solutions

1) 24 is the smallest number with 8 positive factors: 1, 2, 3, 4, 6, 8, 12, 24.

24

2) Use algebra. Write the problem as an equation and solve for x:

$$2x = x/2 + 21$$
$$4x = x + 42$$
$$3x = 42$$
$$x = 14$$

14

3) A carpenter built some decorative tables. Some had three legs and some had 4 legs. When he finished he counted 49 tables and 169 legs. To find out how many 4-legged tables he built make a table or use algebra.

Table solution: To make a table, start with about half of each type of table and see if the number of 4 legged tables needs to be increased or decreased. Since the first scenario yields 171 legs, we need to decrease the number of 4 legged tables so that the number of legs goes down.

3 legged tables	25	26	27
number of legs	75	78	81
4 legged tables	24	23	**22**
number of legs	96	92	88
number of tables (49)	49	49	49
number of legs (169)	171	170	169

Algebra solution: Let t be the number tripod tables and s be the number of square tables. If each tripod table has 3 legs and each square table has 4 legs and there are 169 total legs, then 3t + 4s = 169. There are 49 tables so t + s = 49. To find the number of 4 legged tables (s), isolate the variable s by multiplying the second equation by 3 and subtracting it from the first equation.

$$3t + 4s = 169$$
$$-\ \underline{3t + 3s = 147}$$
$$s = 22 \text{ four-legged tables}$$

22

PERENNIAL MATH — TEST #4

February 2018 Advanced Level Solutions

4) To find how many total balls will you use to make a pyramid with 12 layers, you simply add up the layers which are all squares. Therefore, 1 + 4 + 9 + 16 + 25 + 36 + 49 + 64 + 81 + 100 + 121 + 144 = 650. It will take 650 balls to build 12 layers on the pyramid.

650

5) Use the formula for average rate = distance ÷ time. The total trip is 150 miles each way for a total of distance of 300 miles. Since the average rate is 60 mph, then 300miles/60 mph = 5 hours total. The trip to the coliseum is 150 miles/50 mph = 3 hours. Therefore, 5 hours – 3 hours = 2 hours coming home. 150 miles/2 hours = 75 mph. So they drove 75 mph coming home.

	Rate	Time	Distance
Going	50	3	150
Return	X	2	150
Round Trip	60	5	300

75

6) Work backwards. Since all boys had 16 marbles at the end start there and go in reverse order. Eric had 26 marbles at the beginning.

Step 4. Each ends with 16
Step 3. Trevor gave Eric and Craig as many marbles as each had.
Step 2. Craig then gave Eric and Trevor as many marbles as each had.
Step 1. Eric gave Craig and Trevor as many marbles as each had.

	Eric	Craig	Trevor
End (Step 4)	16	16	16
Step 3	8	8	32
Step 2	4	28	16
Step 1	26	14	8

Therefore, Eric had 26 marbles at the beginning.

26

Name________________________________

PERENNIAL MATH

January 2018 Advanced Level

TEST #1

1) What is the only number between 40 and 50 that has an odd number of factors?	
2) There are 12 candidates in a class election. The winner will be the president, and the runner-up will be the vice president. How many different combinations of president and vice president are possible?	
3) Mary, Debbie, and Jill were the top three winners in bingo. They shared a cash prize in the ratio 2:3:4. Jill received $4 more than Mary. What was the amount of the cash prize?	
4) Find the area of the shaded part. Use 3.14 for π. 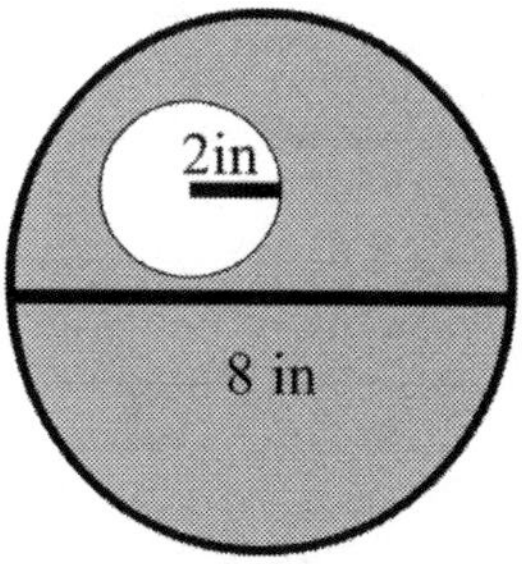	
5) Connie spent two thirds of her money. She lost one-eighth of the remainder and then had $14 left. With how much money did she have at the start?	
6) Arnold has a lot of pennies, nickels, dimes, and quarters in his piggy bank. How many ways can he make 67 cents if he must have at least one of each type of coin?	

PERENNIAL MATH

TEST # 1

January 2018 Advanced Level Answers

1) What is the only number between 40 and 50 that has an odd number of factors?	49
2) There are 12 candidates in a class election. The winner will be the president, and the runner-up will be the vice president. How many different combinations of president and vice president are possible?	132
3) Mary, Debbie, and Jill were the top three winners in bingo. They shared a cash prize in the ratio 2:3:4. Jill received $4 more than Mary. What was the amount of the cash prize?	18
4) Find the area of the shaded part. Use 3.14 for π. 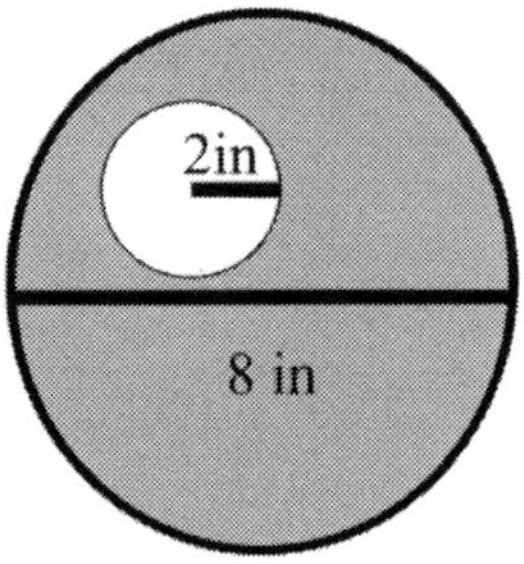	37.68
5) Connie spent two thirds of her money. She lost one-eighth of the remainder and then had $14 left. With how much money did she have at the start?	48
6) Arnold has a lot of pennies, nickels, dimes, and quarters in his piggy bank. How many ways can he make 67 cents if he must have at least one of each type of coin?	13

PERENNIAL MATH

TEST # 1

January 2018 Advanced Level Solutions

1) The only number between 40 and 50 that has an odd number of factors is 49 whose factors are 1, 7, and 49. Composite numbers and prime numbers have an even number of factors unless they are a square.

49

2) There are 12 candidates in a class election. The winner will be the president, and the runner-up will be the vice-president. Once a candidate is picked then they cannot be picked for the other position. So to choose president, there are 12 choices, then once someone is selected, that leaves 11 people remaining for vice-president. So 12 x 11 = 132 possible different combinations of president and vice-president.

132

3) The ratio of the cash prize for Mary, Debbie, and Jill was 2:3:4. Jill received \$4 more than Mary which is 2 more portions. Each portion is worth \$4/2 = \$2. There are a total of 2 + 3 + 4 = 9 portions. 9 x \$2 = \$18. The amount of the cash prize was \$18.

18

4) To find the area of the shaded part you must find the area of the larger circle and subtract out the area of the smaller circle. The area of the larger circle is πr^2 or $\pi(4in)^2 = \pi \cdot 16in^2$. The area of the smaller circle is $\pi(2in)^2 = \pi \cdot 4in^2$. So, subtracting the smaller circle's area from the large circle area we have:

$$(\pi \cdot 16in^2) - (\pi \cdot 4in^2) = \pi \cdot 12in^2 \text{ or } 3.14 \cdot 12in^2 = 37.68in^2$$

37.68

PERENNIAL MATH

TEST # 1

January 2018 Advanced Level Solutions

5) Connie spent two thirds of her money. She lost one-eighth of the remainder and then had $14 left. To find out how much money she started with draw a diagram or work backwards. If you draw out the problem, then find the amount for each portion of the last third (8 portions). $14/7 portions = $2 per portion. Each third of the money is worth $2 x 8 portions = $16, so the total amount she started with is 3 x $16 = $48.

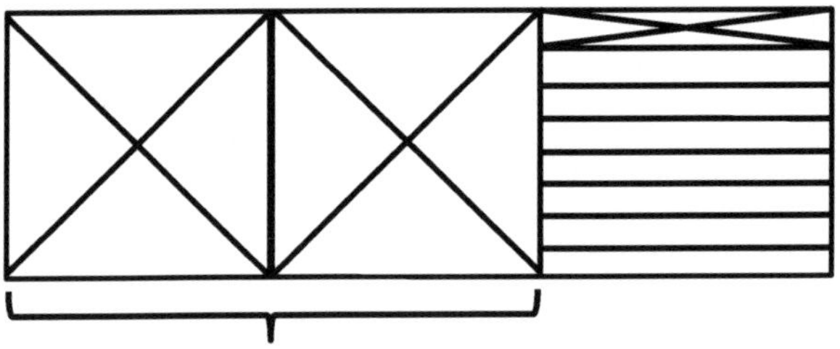

48

6) Make a table to find the different coin combinations that can be made that add up to 67 cents and that have at least each coin used once. There are 13 total combinations that fit the requirements.

Quarters	Dimes	Nickels	Pennies	0.67
2	1	1	2	0.67
1	3	2	2	0.67
1	3	1	7	0.67
1	2	4	2	0.67
1	2	3	7	0.67
1	2	2	12	0.67
1	2	1	17	0.67
1	1	6	2	0.67
1	1	5	7	0.67
1	1	4	12	0.67
1	1	3	17	0.67
1	1	2	22	0.67
1	1	1	27	0.67

13

Name________________________________

PERENNIAL MATH

TEST #2

February 2018 Advanced Level

1) What is the largest two-digit number that is a product of two prime numbers?	
2) Find A • B + C if the following: A = the number of prime numbers less than 80 B = smallest 1 digit prime C = mean of A and B	
3) This rectangular prism has a volume of 288 in². How many square inches are in the surface area of this rectangular prism? 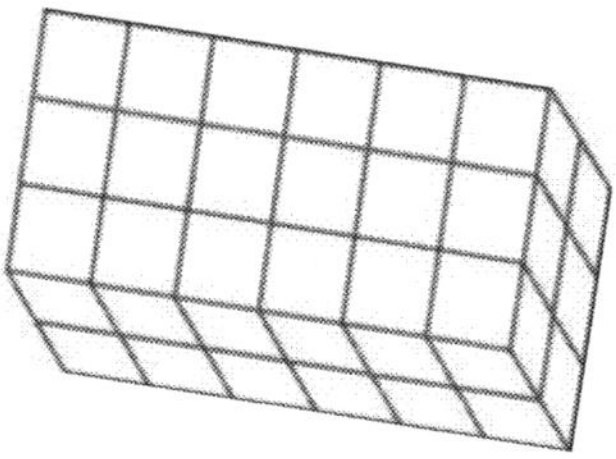	
4) If x ☺ y = x ÷ y + 2, then what is (12 ☺ 6) ☺ 2?	
5) Of the 100 students in the 8th grade, 28 took PE, 31 took Band, 42 took Art, 9 took PE and Band, 10 took PE and Art, 6 took Band and Art, and 4 took all three subjects. How many students took none of the three subjects?	
6) The ages of three cousins are consecutive even integers. Three times the age of the youngest cousin exceeds the oldest cousin's age by 40 years. What is the age of the oldest cousin?	

PERENNIAL MATH

TEST #2

February 2018 Advanced Level Answers

1) What is the largest two-digit number that is a product of two prime numbers?	**95**
2) Find A ▪ B + C if the following: A = the number of prime numbers less than 80 B = smallest 1 digit prime C = mean of A and B	**56**
3) This rectangular prism has a volume of 288 in². How many square inches are in the surface area of this rectangular prism? 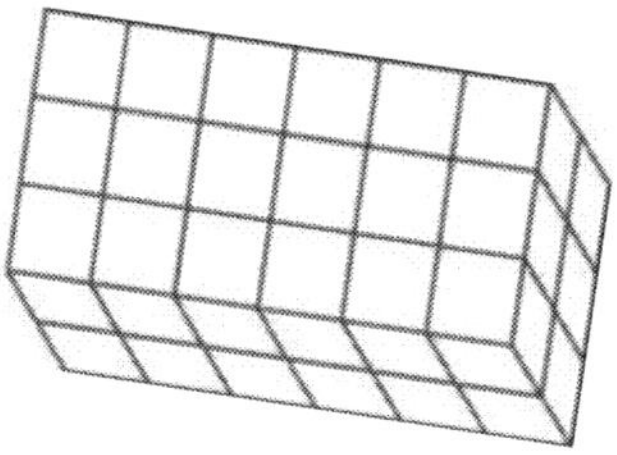	**288**
4) If x ☺ y = x ÷ y + 2, then what is (12 ☺ 6) ☺ 2?	**4**
5) Of the 100 students in the 8th grade, 28 took PE, 31 took Band, 42 took Art, 9 took PE and Band, 10 took PE and Art, 6 took Band and Art, and 4 took all three subjects. How many students took none of the three subjects?	**20**
6) The ages of three cousins are consecutive even integers. Three times the age of the youngest cousin exceeds the oldest cousin's age by 40 years. What is the age of the oldest cousin?	**26**

PERENNIAL MATH

TEST # 2

February 2018 Advanced Level Solutions

1) The largest two-digit number that is a product of two prime numbers is 95. It's factors are 5 and 19.

95

2) Find A ▪ B + C if A = the number of prime numbers less than 80, B = smallest 1 digit prime and C = mean of A and B. To solve for A list and count up the prime numbers less than 80: 2, 3, 5, 7, 11, 13, 17, 19, 23, 29, 31, 37, 41, 43, 47, 53, 59, 61, 67, 71, 73, 79. There are 22 numbers, so A = 22. The smallest 1 digit prime is the number 2, so B = 2. The mean of A and B = (22 + 2)/2 = 12, so C = 12.

Therefore, A ▪ B + C = (22 ▪ 2) + 12 = 56

56

3) This rectangular prism has a volume of 288 in^2. To find how many square inches are in the surface area first solve for the sides of the prism. L ▪ W ▪ H = 288 in^2.

$6x \cdot 3x \cdot 2x = 288\ in^2$

$36x^3 = 288\ in^2$

$x^3 = 288\ in^2/36 = 8.$

$x = 2$

Therefore, each side of the small cubes equal 2 in.

Now that we know the length of those small cubes, we can find out the dimensions of the prism. Length is 6(2in) = 12 in. Width is 3(2in) = 6 in. Height is 2(2in) = 4 in.

To solve for the surface area solve for the area of each side and add those up.

2 ▪ (12 in x 6 in) + 2 ▪ (12 in x 4 in) + 2 ▪ (6 in x 4 in) =

2 ▪ ($72in^2$) + 2 ▪ ($48in^2$) + 2 ▪ ($24in^2$) =

$144in^2 + 96in^2 + 48in^2 = 288in^2$

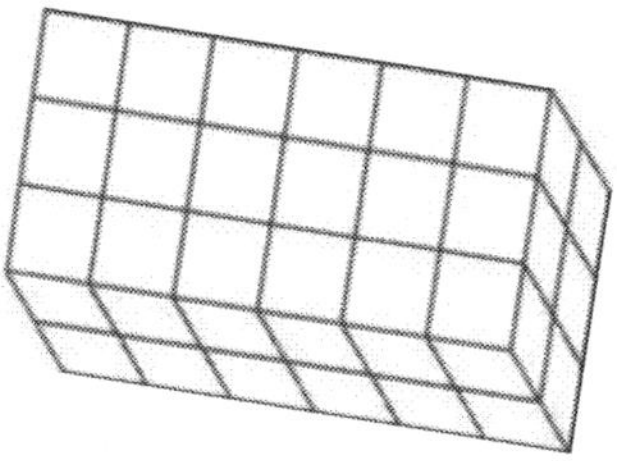

288

PERENNIAL MATH

TEST #2

February 2018 Advanced Level Solutions

4) If x ☺ y = x ÷ y + 2, then what is (12 ☺ 6) ☺ 2?

12 ÷ 6 + 2 = 4

4 ÷ 2 + 2 = 4

4

5) To find out how many students took none of the subjects, use a Venn Diagram. Let x equal the number of students outside the circles. Of the 100 students, 28 took PE, 31 took Band, 42 took Art, 9 took PE and Band, 10 took PE and Art, 6 took Band and Art, and 4 took all three subjects. Put the 4 in the center and work around it to solve for the small regions. 9 took PE and Band so 4 + 5 = 9. 10 took PE and Art and 4 +6 = 10. 6 took Band and Art so 4 + 2= 6. Subtract the overlapping regions from the total of each subject to find the amounts that go inside the circles. For instance PE has 28 total and 28 – (5 + 4 + 6) = 13. Likewise you can solve for Band and Art. The eight regions are 13 + 5 + 4 + 6 + 20 + 2 + 30 + x = 100. 100 – 80 = 20. There were 20 students who took none of the subjects.

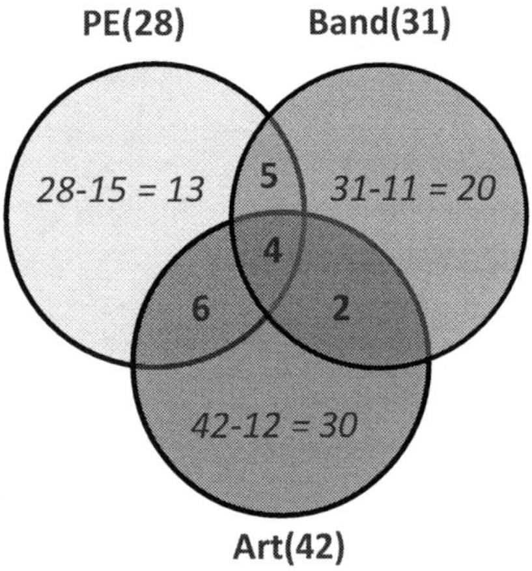

20

6) The ages of three cousins are consecutive even integers. Three times the age of the youngest cousin exceeds the oldest cousin's age by 40 years. Use algebra to find the age of the oldest cousin. The cousins' ages can be represented as: x, (x + 2), (x + 4)

3x = (x + 4) + 40

3x = (x + 44)

2x = 44

x = 22

The cousins' ages are 22, 24, and 26

3(22) = 66

66 – 40 = 26 So, the oldest cousin is 26 years old.

26

Name________________________________

PERENNIAL MATH

March 2018 Advanced Level

TEST #3

1) Find the slope that passes through (0, -2) and (3, 4).	
2) What is the largest three-digit number that leaves a remainder of 1 when it is divided by 5 or by 8?	
3) The Fantasy Playhouse auditorium is renovating its 19 rows of seats. These 19 rows have the same number of seats. It plans to move the seats from the first three rows and add those chairs to the remaining rows which will keep the same number of total seats. The remaining rows would each add an additional 6 seats. How many seats are in the auditorium?	
4) Every positive integer can be written as the sum of four or fewer square numbers: $5 = 2^2 + 1^2 = 4 + 1$ $101 = 4^2 + 6^2 + 7^2 = 16 + 36 + 49$ $170 = 2^2 + 6^2 + 7^2 + 9^2 = 4 + 36 + 49 + 81$ Find three different square numbers that would add up to 83 and then find the difference between the highest and lowest of the three numbers.	
5) The average height of four people is 182 cm. Two of the people are each 174 cm tall, and the third is 186 cm tall. How many centimeters tall is the fourth person?	
6) Kerry and Lynn are riding their bikes around 200 ft circular driveway in the same direction. Kerry travels 5 feet per second and Lynn travels 3 feet per second. How many feet has Lynn ridden when they meet up for the first time?	

PERENNIAL MATH

TEST #3

March 2018 Advanced Level Answers

1) Find the slope that passes through (0, -2) and (3, 4).	**2**
2) What is the largest three-digit number that leaves a remainder of 1 when it is divided by 5 or by 8?	**961**
3) The Fantasy Playhouse auditorium is renovating its 19 rows of seats. These 19 rows have the same number of seats. It plans to move the seats from the first three rows and add those chairs to the remaining rows which will keep the same number of total seats. The remaining rows would each add an additional 6 seats. How many seats are in the auditorium?	**608**
4) Every positive integer can be written as the sum of four or fewer square numbers: $5 = 2^2 + 1^2 = 4 + 1$ $101 = 4^2 + 6^2 + 7^2 = 16 + 36 + 49$ $170 = 2^2 + 6^2 + 7^2 + 9^2 = 4 + 36 + 49 + 81$ Find three different square numbers that would add up to 83 and then find the difference between the highest and lowest of the three numbers.	**40**
5) The average height of four people is 182 cm. Two of the people are each 174 cm tall, and the third is 186 cm tall. How many centimeters tall is the fourth person?	**194**
6) Kerry and Lynn are riding their bikes around 200 ft circular driveway in the same direction. Kerry travels 5 feet per second and Lynn travels 3 feet per second. How many feet has Lynn ridden when they meet up for the first time?	**300**

PERENNIAL MATH

TEST # 3

March 2018 Advanced Level Solutions

1) Find the slope that passes through (0, -2) and (3, 4). Use the slope formula:

$$m = \frac{y_2 - y_1}{x_2 - x_1}$$

$m = \frac{4-(-2)}{3-0}$ = 6/3 = 2

2

2) What is the largest three-digit number that leaves a remainder of 1 when it is divided by 5 or by 8? The least common multiple of 5 and 8 is 40. The greatest 3 digit number that is a multiple of 40 is 960. Therefore, the greatest three-digit number that leaves a remainder of 1 when divided by 5 or 8 is 961.

961

3) The Fantasy Playhouse auditorium is renovating its 19 rows of seats. It plans to move the seats from the first three rows and add those chairs to the remaining rows which will keep the same number of seats. The remaining rows would each add an additional 6 seats. How many seats are in the auditorium? If you draw this out, you will see there are 16 remaining rows and if you add 6 additional seats, that is 16 x 6 = 96 seats that were moved. 96 divided by the 3 rows is 32 seats in each row. Since there were 19 original rows, then 19 x 32 = 608 seats in the auditorium.

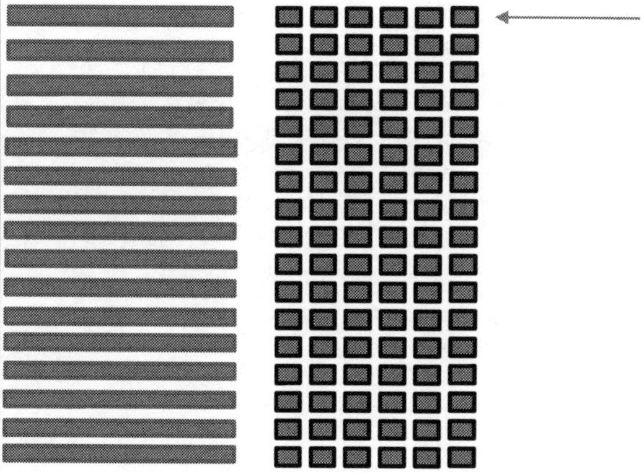

6 seats added to the remaining 16 rows = 96 seats from the 3 rows removed - which means there were 32 seats in each row. 19 rows x 32 seats = 608 seats.

608

PERENNIAL MATH

TEST #3

March 2018 Advanced Level Solutions

4) Every positive integer can be written as the sum of four or fewer different square numbers:

$$5 = 2^2 + 1^2$$
$$101 = 4^2 + 6^2 + 7^2$$
$$170 = 2^2 + 6^2 + 7^2 + 9^2$$

Find three different square numbers that would add up to 83 and then find the difference between the highest and lowest of the three numbers. Square numbers less than 83 are: 1, 4, 9, 16, 25, 36, 49, 64, 81.

$83 = 3^2 + 5^2 + 7^2 = 9 + 25 + 49$

49 - 9 = 40

40

5) The average height of four people is 182 cm. Two of the people are each 174 cm tall, and the third is 186 cm tall. To find the height of the fourth person use the formula for averages and solve for what the missing amount is.

Average = sum of all items/number of items.

$$182 = (174 + 174 + 186 + x) / 4$$
$$728 = 174 + 174 + 186 + x$$
$$728 – 534 = x$$
$$194 = x$$

194

6) Kerry and Lynn are riding their bikes around 200 ft circular driveway in the same direction. Kerry travels 5 feet per second and Lynn travels 3 feet per second. How many feet has Lynn ridden when they meet up for the first time?

This is a distance/rate/time problem. Since it takes Kerry just 200 ft /5 fps = 40 seconds per lap, let's calculate the number of laps the children ride for an arbitrary 10 minutes or 600 seconds using $d = rt$.

Kerry 5 fps x 600 sec = 3000 ft or 3000 ft / 200 ft = 15 laps

Lynn 3 fps x 600 sec = 1800 ft or 1800 ft / 200 ft = 9 laps

So, Kerry meets up with Lynn 15 – 9 = 6 times during that 10 minutes or once every 600 sec/6 = 100 seconds. Since Lynn's speed is 3 fps, then she will have travelled 3 fps x 100 sec or 300 feet before Kerry passes her for the first time.

300

Name________________________________

PERENNIAL MATH

TEST #4

April 2018 Advanced Level

1) What is the sum of the GCF and LCM for 9, 12, and 24?

2) There are 12 monkeys in the zoo. Congo eats more bananas than each of the other monkeys every day. Every monkey eats a prime number of bananas including Congo. Today 57 bananas were eaten. What is the least number of bananas that Congo could have eaten today?

3) If $x \heartsuit y = 3x + y^3$, then evaluate $(2 \heartsuit 5) \heartsuit 3$.

4) The area of right triangle XYZ is 1/6 that of square ABCD. The length of the side of XYZ are 6, 8, and 10 units. What is the area of square ABCD?

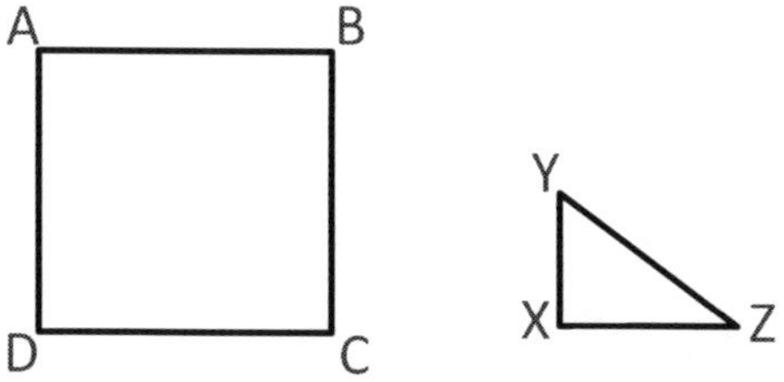

5) Trent read 24° on the thermometer at 7 a.m. He read the thermometer for 4 more hours. By the hour, the temperature rose 3°, dropped 1°, rose 4°, and dropped 2°. What will the temperature have to rise to the next hour for the average temperature to be 28°?

6) This spinner is numbered 1-6. You spin the spinner twice. What is P(greater than 2, odd)? Write your answer as a percentage. P(greater than 2, odd) = ___% (*round your percentage to the nearest hundredth or 2 decimal places*)

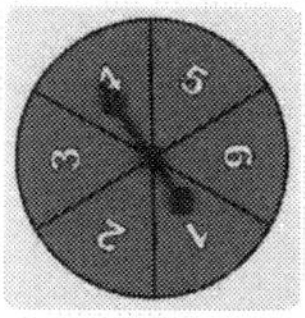

PERENNIAL MATH

TEST #4

April 2018 Advanced Level Answers

1) What is the sum of the GCF and LCM for 9, 12, and 24?	**75**
2) There are 12 monkeys in the zoo. Congo eats more bananas than each of the other monkeys every day. Every monkey eats a prime number of bananas including Congo. Today 57 bananas were eaten. What is the least number of bananas that Congo could have eaten today?	**7**
3) If x ♥ y = $3x + y^3$, then evaluate (2 ♥ 5) ♥ 3.	**420**
4) The area of right triangle XYZ is 1/6 that of square ABCD. The length of the side of XYZ are 6, 8, and 10 units. What is the area of square ABCD? 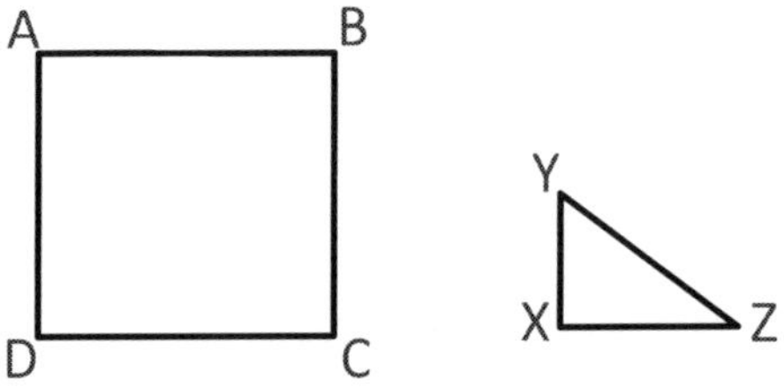	**144**
5) Trent read 24° on the thermometer at 7 a.m. He read the thermometer for 4 more hours. By the hour, the temperature rose 3°, dropped 1°, rose 4°, and dropped 2°. What will the temperature have to rise to the next hour for the average temperature to be 28°?	**33**
6) This spinner is numbered 1-6. You spin the spinner twice. What is P(greater than 2, odd)? Write your answer as a percentage. P(greater than 2, odd) = ___% (*round your percentage to the nearest hundredth or 2 decimal places*) 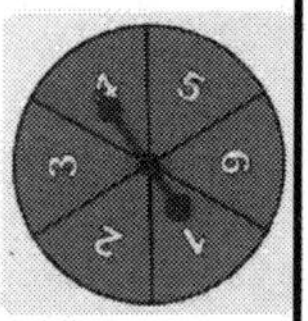	**33.33**

PERENNIAL MATH

TEST #4

April 2018 Advanced Level Solutions

1) The greatest common factor of 9, 15, and 24 is 3. The least common multiple of 9, 12, and 24 is 72. The sum of the GCF and LCM for 9, 15, and 24 is 3 + 72 = 75

75

2) There are 12 monkeys in the zoo. Congo eats more bananas than each of the other monkeys every day. Every monkey eats a prime number of bananas including Congo. Today 57 bananas were eaten. What is the least number of bananas that Congo could have eaten today?
Approach this by asking what is the lowest possible prime number that Congo can have, but it must be higher than all the other monkey. He cannot have 5 because that would only be 33 bananas for all the other monkeys since the most amount each of the other 11 could have would be 3. If he had 7 bananas, the other monkeys would have to have a sum of 50, which is possible since there are 11 other monkeys who could have some combination of 5, 3, and 2 bananas each. So they could have eaten 2 + 3 + 5 + 5 + 5 + 5 + 5 + 5 + 5 + 5 + 5 = 7. 50 + 7 = 57. Congo ate 7 bananas today.

7

3) If x ♥ y = $3x + y^3$, then evaluate (2 ♥ 5) ♥ 3.
Solve for the parentheses first:
If x ♥ y = $3x + y^3$, then evaluate (2 ♥ 5) = $3(2) + 5^3 = 6 + 125 = 131$
Now solve 131 ♥ 3 = $3(131) + 3^3 = 393 + 27 = 420$.

420

PERENNIAL MATH

TEST # 4

April 2018 Advanced Level Solutions

4) The area of right triangle XYZ is 1/6 that of square ABCD. The length of the side of XYZ are 6, 8, and 10 units. What is the area of square ABCD?

The area of the right triangle XYZ is
b ▪ h/2
8 ▪ 6 / 2 = 24 square units.
Since XYZ is 1/6 of ABCD then multiply by 6 to find the area of ABCD.
24 units^2 x 6 = 144 units^2

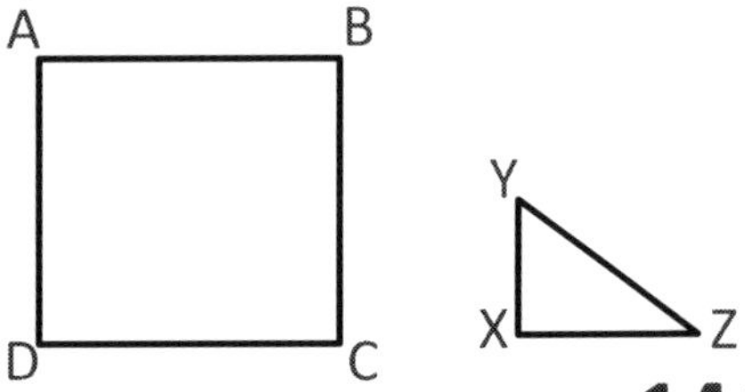

144

5) Trent read 24° on the thermometer at 7 a.m. He read the thermometer for 4 more hours. By the hour, the temperature rose 3°, dropped 1°, rose 4°, and dropped 2°. What will the temperature have to rise to the next hour for the average temperature to be 28°? Make a chart to list the temperature change each hour. Then solve for the missing temperature x.

Time	Temperature
7:00	24
8:00	27
9:00	26
10:00	30
11:00	28
12:00	x

Average = sum of items/number of items

$$(24 + 27 + 26 + 30 + 28 + x) /6 = 28$$
$$135 + x = 168$$
$$x = 33$$

The temperature will need to rise to 33°.

33

6) Two events are independent if the outcome of the first event does not affect the outcome of the second event. The events greater than 2 and odd are independent. The first spin does not affect the second spin. Find P(greater than 2). The spinner has 6 equal sections, numbered 1-6. The numbers greater than 2 are 3, 4, 5, and 6. There are 4 numbers greater than 2. The events greater than 2 and odd are independent. The first spin does not affect the second spin. Find P(greater than 2). P(greater than 2) = 4/6 or 2/3. Find P(odd). The odd numbers are 1, 3, and 5. There are 3 odd numbers. P(odd) = 3/6 or 1/2. Find P(greater than 2, odd). P(greater than 2, odd) = P(greater than 2) x P(odd) = 2/3 × 1/2 = 2/6 or 1/3. Write your answer as a decimal and then convert it to a percentage with 2 decimal points.
1/3 = 0.3333 = 33.33%

33.33%